A COMPLETE INTRODUCTION TO

CICHLIDS

CO-011

Microgeophagus ramirezi *may be a bit difficult to keep but it seems well worth the trouble. Photo by H.-J. Richter.*

One of the most popular of the cichlids is the Angelfish, Pterophyllum scalare. This is a domestic variety called the Marbled Angelfish. Photo by Andre Roth.

A COMPLETE INTRODUCTION TO
CICHLIDS

The Convict Cichlid (Cichlasoma nigrofasciatum) is hardy and easy to breed. Its common name is derived from the striped pattern. Photo by Arend van den Nieuwenhuizen.

A COMPLETE INTRODUCTION TO

CICHLIDS

COMPLETELY ILLUSTRATED IN FULL COLOR

This male Pseudotropheus zebra *has a female pattern but with blue highlights. It is called a marmalade cat and commands a high price. Photo by Dr. Harry Grier.*

—Dr. Robert J. Goldstein—

Distributed in the UNITED STATES by T.F.H. Publications, Inc., 211 West Sylvania
Avenue, Neptune City, NJ 07753; in CANADA to the Pet Trade by H & L Pet Sup-
plies Inc., 27 Kingston Crescent, Kitchener, Ontario N2B 2T6; Rolf C. Hagen Ltd.,
3225 Sartelon Street, Montreal 382 Quebec; in CANADA to the Book Trade by
Macmillan of Canada (A Division of Canada Publishing Corporation), 164 Com-
mander Boulevard, Agincourt, Ontario M1S 3C7; in ENGLAND by T.F.H. Publica-
tions Limited, 4 Kier Park, Ascot, Berkshire SL5 7DS; in AUSTRALIA AND THE
SOUTH PACIFIC by T.F.H. (Australia) Pty. Ltd., Box 149, Brookvale 2100 N.S.W.,
Australia; in NEW ZEALAND by Ross Haines & Son, Ltd., 18 Monmouth Street,
Grey Lynn, Auckland 2 New Zealand; in SINGAPORE AND MALAYSIA by MPH
Distributors (S) Pte., Ltd., 601 Sims Drive, #03/07/21, Singapore 1438; in the
PHILIPPINES by Bio-Research, 5 Lippay Street, San Lorenzo Village, Makati Rizal;
in SOUTH AFRICA by Multipet Pty. Ltd., 30 Turners Avenue, Durban 4001. Pub-
lished by T.F.H. Publications Inc. Manufactured in the United States of America
by T.F.H. Publications, Inc.

Contents

Introduction

Cichlids have long been the most popular egg-laying fishes in the aquarium hobby. Their personality, diversity, coloration, and size have appealed to generations of aquarists. How many of us remember that trip to the home of the serious aquarist, where we first saw the layout, productivity, and magnificent specimens produced by the skilled breeder? The sight of countless young schooling past the air stone or enveloping the parents in a cloud has always thrilled the novice and no doubt always will. And one is always more and more respectful of the proud aquarist who displays his giant pairs of angels, Dempseys, ports, or oscars. We never quite figure out how he has raised such huge fish in such small quarters. And no matter how old we get or how

long in the hobby, there is always that magnificent pair of something to awe us even more.

In recent years the aquarium hobby has been experiencing ever higher peaks of knowledge and activity, and cichlids have figured handsomely in practically every aspect of aquariology. Indeed, the impetus for the opening up of the African aquarium market on a scale never before seen was largely the result of pressure exercised by serious aquarists seeking new and beautiful cichlids.

The Oscar (Astronotus ocellatus) has won over countless aquarists with its "intelligence" and "friendliness". Photo by Andre Roth.

Apistogramma agassizi *is one of those fortunate fishes that has not had its name changed—yet. Photo by H.-J. Richter.*

For years, cichlidophiles have classified their fishes into four categories: the African, the Asian, the American big ones, and American dwarf cichlids. The scientist, however, is not concerned with any aquarium classification. His interest is in lines of evolution and the mechanics of the formation of species through time in response to an ever-changing environment, an environment not merely physical and climatic, but biological as well. Competition for food and space and adaptations to temperature changes are only part of the picture. New predators and new diseases take their toll not only of our cichlids, but of their predators and prey as well. And the pressures of this complex environment may result in animals of only distant relationship both changing, through the eons, toward similar end-points. This is convergent evolution, and it can play havoc with any scheme of classification based on looks alone. The scientist is concerned with the evidence of the fossil record, of development, of chemistry of the body, and of the hereditary material as visualized in the

12

chromosomes. In practice, however, much of this evidence is not yet available, difficult to come by, or just not worth the effort considering the difficulty of the problem. The ichthyologist, therefore, is primarily concerned with those things that can be well-preserved, and consequently studies mostly the scales, bones, teeth, musculature, and geographic distribution of

Lamprologus leleupi comes from Lake Tanganyika. Newly imported individuals of this morph are bright yellow. Succeeding generations are less colorful.

This species, on the other hand, is currently being bounced from one genus to another. It may be neither Pelmatochromis thomasi *nor* Hemichromis thomasi, *the two currently used combinations. Photo by H. Hansen.*

species (zoogeography).

Put in simple tabular form, the cichlids bear the following relationship to the rest of the bony fishes:

TELEOSTS
Superorder Acanthopterygii
Order Perciformes
Suborder Percoidei
Family Cichlidae

The cichlids are only one family in a huge order, but are distinguished by a set of characteristics. There is only one nostril on each side of the head, instead of two. The marine family Pomacentridae is quite similar, but differs from the cichlids in possessing a bony plate near the orbit (the eye socket)—a bone the cichlids lack. In most cases, the lateral line is divided into an upper long lateral line forward and a lower short lateral line toward the rear. The scales along the lateral line are usually counted in delineating species, and this divided lateral line has been "scale-counted" in different ways by different ichthyologists, resulting in some confusion of which species is which. Cichlids have both spiny and soft dorsal and anal fins. That is to say, the forward rays of these fins are often

Cichlid teeth are often used as an aid to identification. These tricuspid teeth with dark tips are from Petrotilapia tridentiger. *Photo by Dr. Herbert R. Axelrod.*

unbranched (spiny), and the rear ones are branched (soft rays). Again, the terminal rays of these fins are often very short, and some workers have counted the little ones and some have ignored them. The teeth of the cichlids are very variable, and often of importance at the genus level. Cichlid teeth are almost always brown at the tips. The teeth may be simple (conical), very sharp, flattened for crushing, divided (bicuspid, tricuspid), few and far between, densely crowded, and in one or many rows in one or both jaws. They may point forward, straight up and down, or backward, be long or short, extend along the sides of the jaws, or extend into the pharynx (almost always). There may be all the same type of teeth in the jaws, or different types in the front and along the sides. The pharyngeal teeth may vary in size, shape, and distribution. The distribution of scales on the head may vary, as may the size and design of the scales. The shape of the head varies from concave to straight to convex, above and below, with or without indentations in the profile. The body may be bass-shaped, disc-shaped, or bizarre as in the angel fishes. The variation in shape is matched by extraordinary variation in markings and colors, and just leafing through this book will give you an idea of the variation in this large

Color pattern and shape are heavily relied on for identification of living aquarium specimens. This is the wide band variety of Tropheus duboisi *from Lake Tanganyika. Photo by H.-J. Richter.*

family. Behavior is important in speciation and may also be of value in classification.

What has classification got to do with the aquarist? You and I know that we are never satisfied with even the generic identification of our fish; we want to know the exact species. This can only be done by a careful examination of the number of rays in the fins and the

scale counts, at the very least. Pictures are a naive approach. You must expect that there are several species that look very much alike, even if only one of them has made the grade as a species illustrated in aquarium textbooks. If a fish's identification is unknown to you and you want it identified, there are certain things you must do. First, preserve it in rubbing

Classification

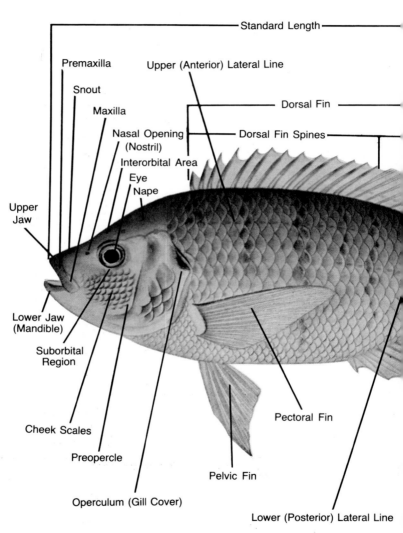

Standard Length

Premaxilla

Upper (Anterior) Lateral Line

Snout

Dorsal Fin

Maxilla

Nasal Opening
(Nostril)

Dorsal Fin Spines

Interorbital Area

Eye

Nape

Upper
Jaw

Lower Jaw
(Mandible)

Suborbital
Region

Cheek Scales

Pectoral Fin

Preopercle

Pelvic Fin

Operculum (Gill Cover)

Lower (Posterior) Lateral Line

*Schematic diagram of a generalized cichlid (*Sarotherodon
mossambicus*).*

alcohol before it has died
and decomposed. The
belly should be slit with a

razor blade to allow the
preservative to get inside
quickly. With a magnifying

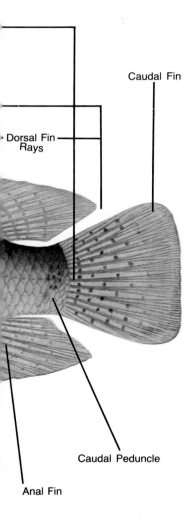

Caudal Fin

Dorsal Fin
Rays

Caudal Peduncle

Anal Fin

soft rays by Arabic numerals. The two types are separated by a stroke, e.g. XII/9. With a clear plastic ruler measure the length of the fish from the tip of the snout to the base of the tail fin; this is the Standard Length. Most other measurements are reported as percentages of standard length, as this is usually rather constant with a species over a range of sizes. You must give the standard length, however, because some species illustrate changes in the percentages with increasing size. The snout length is the distance from the tip of the snout to the front edge of the eye. The head length is the distance from the tip of the snout to the rear edge of the head, as measured by the edge of the gill cover. The peduncle length is measured from the rear of the anal fin to the base of the tail fin. Some other measurable characteristics are more difficult to do with a ruler, but if you've gone this far you have made a legitimate effort. If you cannot identify the fish on the basis of this work, you can now send it off to an ichthyologist, together with a letter stating your measurements, your source of the fish, and its

glass, the aquarist should do his best to count the dorsal, anal, pectoral, and pelvic fin rays. The spiny rays are designated with Roman numerals and the

Classification

coloration in life. Wrap it in a rag soaked in alcohol, place it in a plastic bag, and send it to an ichthyologist. Be sure to write first and make sure the ichthyologist will have time to look.

LIST OF MAJOR CICHLID GENERA INCLUDING SOME SYNONYMS

Acarichthys
Acaronia
Aequidens
Amphilophus
Apistogramma
Asprotilapia
Astatheros
Astatochromis
Astatoreochromis
Astatotilapia
Astronotus
Aulonocara
Aulonocranus
Baiodon
Bathybates
Batrachops
Bayonia
Biotodoma
Biotoecus
Boggiania
Boulengerochromis
Callochromis
Cardiopharynx
Chaetobranchopsis
Chaetobranchus
Chaetolabrus
Chaetostoma
Chilochromis

Chilotilapia
Chromichthys
Chromidotilapia
Cichla
Cichlasoma
Cichlaurus
Clinodon
Cnestrostoma
Coptodon
Corematodus
Crenicara
Crenicichla
Cunningtonia
Cyathochromis
Cyathopharynx
Cynotilapia
Cyphotilapia
Cyrtocara
Dicrossus
Docimodus
Ectodus
Enantiopus
Eretmodus
Erythrichthys
Etroplus
Genyochromis
Geophagus
Gephyrochromis
Gobiocichla
Grammatotria
Gymnogeophagus
Haligenes
Haplochromis
Haplotaxodon
Hemibates
Hemichromis
Hemihaplochromis
Hemitilapia
Herotilapia
Heterochromis
Heterogramma
Heterotilapia
Hoplarchus
Hoplotilapia

18

Hygrogonus
Hypsophrys
Julidochromis
Labeotropheus
Labidochromis
Labrochromis
Lamprologus
Lepidolamprologus
Leptochromis
Leptotilapia
Lestradea
Lethrinops
Limnochromis
Limnotilapia
Lipochromis
Lobochilotes
Macropleurodus
Melanochromis
Melanogenes
Mesonauta
Mesops
Microgaster
Mylochromis
Nandopsis
Nannacara
Nannochromis
Neetroplus
Neochromis
Neotilapia
Ophthalmotilapia
Orthochromis
Otopharynx
Oxylapia
Paracara
Parachromis
Parapetenia
Paratilapia
Paretroplus
Pelmatochromis
Pelvicachromis

Perissodus
Petenia
Petrochromis
Petrotilapia
Plataxoides
Platytaeniodus
Plecodus
Pseudetroplus
Pseudopercis
Pseudoplesiops
Pseudotropheus
Pterophyllum
Ptychochromis
Ptychochromoides
Retroculus
Rhamphochromis
Saraca
Sargochromis
Sarotherodon
Satanoperca
Schubotzia
Serranochromis
Simochromis
Spathodus
Stappersia
Steatocranus
Symphysodon
Tanganicodus
Teleogramma
Telmatochromis
Theraps
Tilapia
Tomocichla
Trematocara
Tropheus
Tylochromis
Uaru
Xenochromis
Xenotilapia

Behavior

Cichlids are interesting fish to watch. The reason, obviously, is because they behave in ways surprisingly familiar and understandable to us. One problem in describing cichlid behavior, however, is that it is considered almost sinful among biologists to ascribe human-type emotions or rational behavior to a lower animal. Instead, biologists tend to look for physical or chemical stimuli to explain

Jaw-locking is common in cichlids. It is seen quite often in pre-spawning situations as shown here with this pair of Jack Dempseys (Cichlasoma octofasciatum).

Some cichlids are extremely aggressive. Genyochromis mento (center of photo) for one is a real battler. Notice that most of these cichlids have ragged fins. Photo by A. Ivanoff.

such behavior. But can't the same thing be said of people? Cichlids, as psychologists have long

known, can learn. Fry may learn to follow a parent and enter a hole (especially in cave brooders and mouthbrooders). Adults learn to recognize a mate. Your fish learn to recognize you when you approach the aquarium with food, whereas frequently strangers will elicit no such response. This is not to belittle the work on seeking physical and chemical stimuli for almost all aspects of behavior. What I am leading to is my reason for explaining and/or describing cichlid behavior in terms acceptable to

20

Several cichlid species have young that feed off body secretions especially produced by the parents. Because of this some young are difficult to raise away from the parents. Shown here is the Brown Discus (Symphysodon aequifasciata*). Photo by Yohei Sakamoto.*

aquarists. If we can overlook the technical explanations when speaking of human behavior, then I feel it is legitimate to do the same when describing the behavior of these very intelligent fishes.

The study of behavior is called ethology. What we usually refer to as instinct may be considered the predicted behavior of a species under certain conditions. Certainly, much of this is inherited, but some of the behavior is learned, sometimes from parents, and sometimes by trial and error. Some fishes can be fooled by raising them with fishes of other species. Later in life they may not recognize their own species. Fishes are also adaptable. In the absence of a suitable mate of their own species, they may mate with a fish of another species. Some behavioral sequences, once started, cannot be turned off. In cichlids, this often translates into the situation wherein you find fish getting ready to spawn in a pet shop. You buy them, take them home and place them in very different water and lighting conditions, and they go ahead and finish what they started. Again, fishes may be frightened by bright lights or sudden movements, but when spawning they are often oblivious to these usually objectionable stimuli.

Many cichlids are pugnacious. The pugnacity is usually rooted in their territoriality. A dwarf cichlid from Central America may take over a square foot area in a corner of the aquarium. Another of the same species may take the other corner, and woe unto him who ventures into the staked-out backyard. But

Geophagus balzanii is a delayed mouthbrooder, picking up the eggs a day or two after they are laid. The fry are released from the mouth (above) in order to forage but dash back to safety (below) at the first sign of danger. Photos by H.-J. Richter.

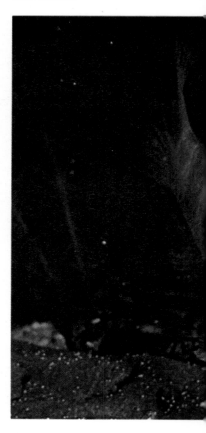

suppose there are less than two square feet of territory? Cichlids will adjust. They will threaten each other by erecting their spiny fins and expanding their gill covers to make them look larger, and they may even resort to chasing and biting their adversaries. But they will adapt. Work with *Nannocara anomala* has shown that the adversaries will eventually settle for a slightly smaller territory, even though they defend the boundaries vigorously. Add a third fish to the tank, keeping the space constant, and they may accept even smaller territories after the usual hassling over who gets how much. Eventually, one can often add enough fish to limited space to destroy their will to defend a territory. This often happens in aquarium shops. A large number of cichlids in one tank in the store seem to get along fine. But take a half dozen home to put in a 20 gallon tank, and before you know it they are fighting over the best rocks and corners.

Many fish are uninterested in maintaining territories except at spawning time. Angels are a good example of this. Also at spawning time,

pairs will share a territory that previously belonged to only one of them. And there may be a change in ownership of the property. The male may stake out the territory, allow the female to enter for spawning, and then be chased out by the female

Many cichlids are fiercely protective of their young. This male Firemouth (Cichlasoma meeki) threatens with expanded gill covers and bright red branchiostegal membranes.

who will then brood the eggs and fry. In the Lake

Malawi (East Africa) "mbuna" group of mouthbrooders, the male defends a territory until a sexually ready female comes along. Both protect the territory during the short spawning period, and then she is chased away from the territory— male cichlids threaten rivals by slowly arching their bodies back and forth, as though they were beating their tails against the enemy. On the other hand, they may carry out the same motions at an eye-blurring speed, and females think this is sexy.

seduced and abandoned! Fish have signals that their own species recognize, and that other species recognize. These signals may be dark markings, bright colors, erection of fins, or other acts of behavior. Many Partners may twitch their heads at each other, and a similar head flick may mean something quite different to a bunch of fry watching their mother. Cichlids may lock jaws, and this usually is a means of fighting. In *Cichlasoma,*

25

mates often do this before spawning, and perhaps this is part of the ritual of forming the pair bond.

We've already talked of threatening behavior. But how does a fish signal that he gives up? There are various ways of signaling appeasement. Some fish show their bellies to the winner. Others lower their heads. And still others change their coloration or pattern to signal "Uncle." Some cichlids will fight until they can hardly be expected to recover from their wounds. Others won't fight at all. Some will vigorously defend their young, and others won't spend any effort at defense. The behavior of some mouthbrooders probably has some easily testable answer. Some

Spawning, egg guarding, and fry tending may entail color changes in the parents. The Etroplus maculatus *tending eggs here has developed a black area on the lower part of its body. Photo by H.-J. Richter.*

Cichlid tanks should usually have some cover in the form of caves, overturned flower pots, etc. Substrate spawners must be provided with a flat surface (angelfishes often spawn on a slate leaned on the side of the tank).

change their minds and go back to eating. Disease or behavior? These are only some of the myriad activities one may expect to see among even the common aquarium cichlids. And among non-aquarium species, there are even more interesting, if not bizarre, kinds of activity. There are cichlids that mimic other cichlids, sneak into a school of these, and bite them to get food. Some eat scales, and some eat eyes. Instinct? In nature, we tend to think of everything as instinctive. But in the aquarium, we often have second thoughts. We get to know our fish as we see them all the time, not some of the time. And attributing these

won't eat at all while brooding, while others will put the eggs down somewhere reasonably safe, catch a quick lunch, and pick up the eggs again. Why do some cichlids, notably discus, suddenly go on hunger strikes, perhaps never again to eat until they die? Yet others suddenly

This female Pseudocrenilabrus multicolor *is showing a lowered gular area which is typical when she is brooding eggs or young. Some mouthbrooders will eat during the brooding period, others will not. Photo by H.-J. Richter.*

activities to the catch-all, "instinct," is just an oversimplification.

One aspect of cichlid lore deserves special comment. It has often been written that cichlids may mate for life. Nonsense! Under most community tank compatible pair; fish that will accept each other frequently and will not fight excessively during non-spawning periods.

The aquarist has used physical stimuli to set off sequences of behavior. Raising the temperature is a commonly employed

Cichlasoma severum *is one of those cichlid species that will spawn repeatedly with the same mate once their compatibility has become established. Photo by R. Zukal.*

conditions, if a choice of mates is available, the partners may switch in subsequent matings. In non-community tanks, if one of a "mated pair" dies, the other partner will often accept a new mate if offered. The "mated pair" concept is a myth. What it really means is a stimulus for triggering angels and discus, and works with some dwarf cichlids as well. Increasing the duration and intensity of light is another such stimulus. Some salt added to the water often triggers increased activity in Malawi cichlids, as does a major water change.

Choosing Cichlids

It is best to select a half dozen young fish, grow them up, and let them choose their own mates. This is a batch of young Lamprologus leleupi. *Photo by H.-J. Richter.*

Most successful cichlid breeders purchase young stock in some quantity, raise the fish under the local water conditions (which may differ from the water their parents spawned in), and allow the fish to choose their own mates. This is very desirable for a number of reasons. To mention just a few: (1) by raising the fish yourself, you can be sure that the adult survivors are well-adapted to your water and foods and feeding schedule; (2) the low cost of young fish enables you to purchase a large number and perhaps get several pairs; (3) you allow the fish to choose compatible partners and avoid unnecessary fights between belligerent strangers; (4) you've made maximal use of the large tank in which you are growing your stock.

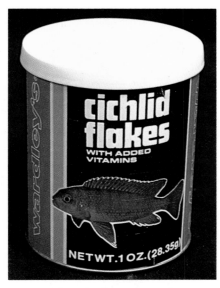

Many foods on the market are designed especially with cichlids in mind. The labels will specifically have the word "cichlids" prominently displayed.

Live natural foods are by far the best fare for your cichlids. Tubificid worms are eagerly taken. These worms usually come in bunches like this.
Photo by Michael Gilroy.

In nature, many species of cichlids are insectivorous, predaceous, or even herbivorous. For example, many of the Lake Malawi "mbuna" complex of mouthbrooders eat only algae in nature. But in the aquarium, almost all cichlids relish a diet of meat. Prepared flake foods are low priced, but should constitute a supplement, not the major part of the diet. The bulk of the diet should consist of a blend of meats. I use raw fish ("ocean perch") as the main ingredient, and to it add some liver (pork, beef, or chicken). One can also add fish roe, cod liver oil, old microworm culture, egg yolk, or anything else he desires. It is a good idea to add some vegetable matter, including mashed spinach or, perhaps, some of your excess duckweed. The entire lot of food is made into a mash in the blender, frozen into cubes in an ice-cube tray, and the cubes thawed before feeding. The value of a mixture is to be found in the biochemical contributions of these various ingredients.

Many cichlid breeders are concerned with maximal growth in minimal time. This is best accomplished in two ways.

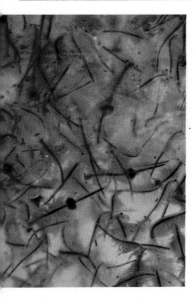

feeding microworms: The tank containing the fry should not have coarse gravel. Bare tanks are best, but fine sand is tolerable. The reason for this is that the fry may attempt to follow their prey into the gravel, become lost and lodged, and die in such set-ups.

If the fry are left with the parents, it is often not necessary (although it is still desirable) to feed live brine shrimp nauplii. The parents will frequently

Another basic live food is brine shrimp. Newly hatched brine shrimp are excellent for newly hatched or very tiny cichlids, adult brine shrimp (shown here) for the larger individuals.

Certain fishes, for example the Discus, provide food for the young through secretion of a specially nutritious material. The young pick this material off the body. Photo by S. Kochetov.

One should feed the fish frequently, at least twice a day and preferably more often. Second, it is useful to leave the light on in the fish room 24 hours a day. This usually will not disturb spawners, and in some instances the long photoperiod may trigger nuptial behavior.

Most cichlid fry can take newly hatched live brine shrimp as soon as they are free-swimming. This is the best food available. Microworms are also useful, especially for the smaller cichlid fry. One important caution about

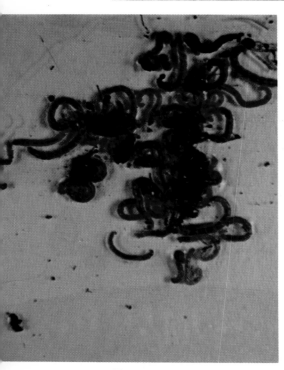

live minnows. The use of minnows should be done cautiously. Some minnows carry monogenetic trematodes ("flukes") on the gills or skin. It is possible to infect some cichlids with the less host-specific parasites on such bait fishes. Feed bait fishes once, and then wait at

Blood worms (chironomid larvae) are avidly accepted by almost every cichlid. These are not as available alive as tubificid worms or brine shrimp but are packaged frozen, freeze dried, etc. Photo by Charles O. Masters.

Hemichromis elongatus thrives on live foods. A balanced diet is as important to fishes as it is to humans.

chew up the adult food into a fine mush and spray it onto the shoal of fry.

Live foods are always best for adult cichlids, but are not always available. Serious cichlid breeders will attempt to grow earthworms in boxes in the backyard, using soil, dead leaves, and stale bread as food. Bait stores supply crickets, mealworms, and

Feeding Cichlids

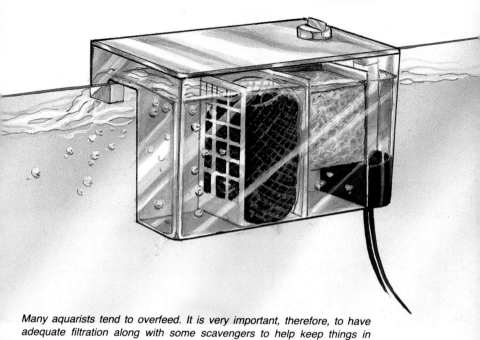

Many aquarists tend to overfeed. It is very important, therefore, to have adequate filtration along with some scavengers to help keep things in balance.

least three weeks before using them again. If in the meantime your fish show signs of scratching themselves or expanded gill covers, discontinue the use of bait minnows from that source and treat the tank with formalin, about four drops to the gallon.

Two types of filter units: a "waterfall" type power filter at left and a canister type power filter at right.

Physical Parameters

Irrespective of the quality of the water in their natural habitats, most cichlids will do quite well in water that is neutral to somewhat alkaline. It should be moderately hard (about 6-10 DH; for ppm multiply by 17.1). And it should be exceedingly clean and well-aerated. There are a number of exceptions to this general rule. Discus do best in soft, slightly acid water, without too vigorous aeration. Some salt is appreciated by many species; for example the cichlids from the large African lakes and species of *Pelvicachromis,* most of which are from coastal areas or somewhat saline pools. Heaters are, in most

Proper lighting is very important. A light situated near the back of the tank (below) shines on the back of the fish leaving the part facing the observer in shadow. If the light is moved toward the front of the tank (right) the light reflects off the front of the fish allowing their full colors to be properly displayed.

cases, unnecessary and carry the risk of a defective thermostat cooking your fishes. Room temperature in a house, if comfortable for you, is comfortable for your fishes. If the fish room is a garage or other unheated room, aquarium heaters should not be used; get an electric floor heater with fan drive. They will never cook fish. A very few species require somewhat elevated temperatures to induce the pre-nuptial play and start the spawning sequence. Heaters may be used in these aquaria. In fact, the only times I use aquarium heaters are for inducing spawning of some fishes where all else has failed, or

Physical Parameters

I am anxious to get rid of a case of *Ichthyophthirius* ("ich") infestation of the tank.

A large tank size to fish size ratio is important to

The tank decor must suit the inhabitants. A well-planted tank such as this one would be fine for some cichlids but others might look at this set-up as a salad course while still others would simply dig or tear everything up leaving the tank in shambles.

growing the fish, but usually not necessary for breeding pairs. Many angel breeders keep their spawning pairs in 6 to 10-gallon aquaria, bare, with good aeration. Larger aquaria and refrigerator liners may be used for growth.

The Notebook

A notebook is a rarely used, yet essential, part of every fish room, whether you're a cichlid fan or any other kind of aquarist. A simple composition book or spiral bound notebook is adequate, and the habit of using it should be cultivated. Whenever I get a good species, I begin a page with the name of the fish, the date, and my source. I also list my source's source if this is known. Any interesting behavior is made note of, as is the size of the fish when I get them, and their size at the first spawning, etc. One should list references to articles on

the fish in national publications, as well as the pages in your several textbooks where these fishes are discussed. I always put down the dates of my observations, including spawning, hatching, free-swimming, maturity, etc., as well as how I handled the eggs, in what way the fish spawned

The notebook is essential in keeping information about breeding. Time of spawning, number of eggs, time it took them to hatch, and time before the fry become free-swimming and ready for food are only a few items that should be recorded. Photo of Aequidens vittatus *by Rainer Stawikowski.*

*For those aquarists that want to breed special strains a notebook will be necessary to indicate the results of various crossings. For example, how many Angelfish (*Pterophyllum scalare*) in this cross were simply marble Angels and how many were Marble Veil Angels. Photo by H. Smith.*

should be also written down in the appropriate section of your notebook. The number, size, and color and shape of the eggs should be recorded, as well as the number of fish raised. Colors and patterns of the parents are interesting in themselves, and often shed light on what to expect from related species.

(on a rock, in a flower pot, etc.), how the parents behaved if left with the fry, whether I used methylene blue or acriflavine dye, and everything else which *might* be important some day. The memory is an unacceptable substitute for careful notes. If you become aware of scientific literature, these references

Aequidens vittatus is rarely seen for sale in this country. It is important that those aquarists who breed them keep a notebook and report their findings to others who might eventually get a pair. Photo by R. Stawikowski.

As Asia is home to only two cichlids of interest, we should dispense with them at once. These are the green chromide, *Etroplus suratensis*, and the orange chromide, *E. maculatus*.

Very little is known of the green chromide as an aquarium fish, and it behaviorally studied by Ward and Barlow in 1967, from whom most of the following information has been taken. About a week after the pair have accepted each other, spawning takes place on a rock or in a cave. The eggs hang from threads, and

apparently cannot be maintained in the freshwater aquarium for long periods. It is probably only rarely imported.

The orange chromide has long been a favorite aquarium fish, being easy to keep and breed, and quite distinctive in shape and coloration. It has been

Only two cichlids are endemic to Asia (actually India and Sri Lanka). This one is the Orange Chromide (Etroplus maculatus). It is generally available to aquarists and breeds readily if proper conditions are available. Photo by H.-J. Richter.

both parents care for the brood, removing dead eggs and fanning the

The second species, Etroplus suratensis, *is a member of the same genus and is called the Green Chromide by aquarists. This species is only rarely available.*

spawn. Pits are dug in the gravel in the meantime. Hatching occurs on the third day, and the wrigglers are moved to the pits for 5 or 6 days, after which they become free-swimming. The fry remain in a tight school around one or both parents and are seen to glance off the sides of the adult fish. During this glancing activity, the fry are feeding on mucus. The number of mucus cells in the parental skin during this period has increased by about 34%. Because the parental slime provides important nutrition during the first critical week (as well as thereafter, when it is no longer critical), the largest batches are raised if the parents are allowed to keep the eggs. This is also true of the South American discus fish

Some aquarists create a background by using pieces of slate in the manner shown.

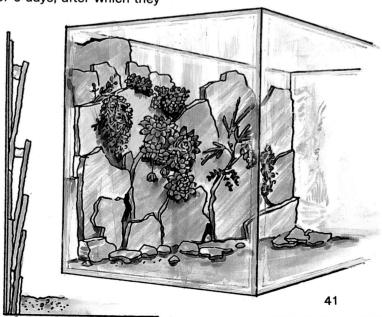

Many cichlids become so aggressive toward their mates that a divider must be employed to prevent one from killing the other.

flicking their dark pelvic fins.

Orange chromides do well in alkaline water (pH 7.5-8.5), with some marine salts added to their water, and a temperature just slightly above room temperature (about 82°F). Much of the present aquarium stock is commercially bred in Florida ponds.

Below: A yellow morph of the Orange Chromide. This is the most popular variety. Photo by H.-J. Richter. Above right: The Green Chromide grows quite large (to 40cm). These are relatively young. Photo by Dr. Herbert R. Axelrod.

(Symphysodon), in which very few aquarists have raised any fish at all when the eggs were removed from the parents. After about three weeks the parents lose interest in the fry and may spawn again. The coloration and patterns of the parent fish are not consistent. Some pairs may turn very yellow when in breeding condition, the coloration reaching a peak about a week after spawning when the fry are about to become free-swimming. They signal danger by

The Cichlids of Africa

The Genus *Tilapia*

Tilapia species are large fishes with limited appeal to aquarists today. They are generally difficult to identify, if only because there are so many species which aquarists have never seen. They occur all over Africa, with most species found in the eastern lakes. They are lake (lacustrine) and river (riverine) fishes, frequently found in open waters feeding on plant material. Six species are bottom spawners (*T. sparrmanii, T. mariae, T. rendalli, T. zillii, T. tholloni* and *T. guinasana).* All other species are mouthbrooders. Among the mouthbrooding species, only one is a paternal mouthbrooder (the male incubates the eggs); this is the blackchin, *T. melanotheron* (also known by the presently invalidated names, *T.*

One of the more unusually patterned species of Tilapia *is this T. buttikoferi. It was an instant favorite with aquarists. Photo by E. Roloff.*

Tilapia tholloni *is one of the bottom spawners. As such it would not have its name changed to* Sarotherodon tholloni *if the recent splitting of the group by Trewavas is accepted. Photo by Dr. Robert J. Goldstein.*

macrocephala, T. microcephala, and *T. dolloi).* All the other are maternal brooders.

Tilapia species are abundant in African lakes.

A number of species, especially the riverine forms, are widely distributed throughout

Africa, whereas certain lakes, especially L. Malawi, are rich in *endemic* forms (species which occur there and nowhere else).

Technically, *Tilapia* Smith, 1840 is characterized by its pharyngeal structure, cycloid scales, an outer series of bicuspid teeth,and several inner series of tricuspid teeth, although large fish may have some inner conical teeth. Young fish often have an ocellus on the dorsal fin. The lower jaw and/or the belly may be black or red or both or neither. All species of *Tilapia* are predominantly vegetarians, feeding on algae and other plants and especially in open surface waters on the diatoms and other floating algal cells. They should be given some plant material in the aquarium, and they are an excellent control on duckweed. Because of their rapid growth, large size, simple food requirements, and adaptability to many kinds of water (sometimes even marine!), they have been seeded all over the world in fish ponds and rice paddy fields, and thus are an important source of fish protein for many otherwise protein-deficient cultures.

In some cases it has been found that the fish become so successful that they quickly reach breeding age, overpopulate the area, and become stunted from crowding, thus reducing their usefulness as food. To offset this, males of one species are stocked with females of another species. They cannot overpopulate an area as the hybrids are sterile. A beautiful hybrid which develops a rich golden color is now being widely used in pond culture. Unfortunately, it is sterile and thus has no

This is probably Tilapia mossambicus *(or* Sarotherodon mossambicus*) which is a mouthbrooder. Almost all species of* Tilapia *are very hardy and prolific making them a prime candidate for aquaculture. Photo by Dr. Herbert R. Axelrod.*

future as an aquarium fish (although this is fortunate for the people it may feed).

The Genus *Hemichromis*
Hemichromis Peters, 1857,

The members of the genus Hemichromis *are in a confused state at present. The common Jewel Fish, long known under the name* Hemichromis bimaculatus, *may not be that species at all. Photo by H. Hansen.*

47

A bright red Hemichromis, *formerly
known under the name* H.
bimaculatus *II is probably actually* H.
lifalilli. *This female is laying another
string of eggs on a flat stone. Photo
by H.-J. Richter.*

is a small genus containing only a few species. The ones commonly kept in the hobby have been called *H. bimaculatus* Gill, 1862 and *H. fasciatus* Peters, 1857, but according to a recent review *H. bimaculatus* is not the common hobby species and the genus includes several almost identical species that are almost impossible for hobbyists to tell apart. When used in the old sense of a single variable species, *"H. bimaculatus"* is easily recognized; this is the African jewel fish, long known to the hobby, easily kept, and easily bred. *H. bimaculatus* II is considered a separate species. The second species, sometimes referred to as the five spot or banded cichlid, is frequently available, and yet very poorly known. There are a number of reasons for this. First, the species is widely distributed in Africa, from far West Africa eastward to the Congo and Niger Rivers, and thence southward to about Southwest Africa. When a species ranges this widely, it is expected to vary considerably in colors and sometimes in morphological characters as well, and this is the case with *H. fasciatus.* Second, there is some evidence that two species are involved in the various references to *H. fasciatus.* One is indeed this species, and the other (quite similar) species is called *H. elongatus.* The genus *Hemichromis* is distinguished by a pair of large teeth in the center of the jaw.

Hemichromis fasciatus often appears in shops at a trusting size of about an inch. These young fish are often ragged and low in price. They grow quickly and don't seem to stop! An aquarium length of six inches is common, and they get much larger in nature. Spawning occurs at about five inches. The pair are excellent parents

The Five-spot Cichlid (Hemichromis fasciatus) quickly grows to a size of six inches or more. They are substrate spawners that protect the fry until they attain a size of about an inch. Photo by Klaus Paysan.

and will continue protecting very young free-swimming fry against all adversaries. Protection continues until the young reach about an inch. After this time the young become quite vicious toward their own kind and must be given much room.

Spawning is typical of a substrate-breeder. The eggs are laid on a hard surface in a careful

also requires no great effort. The fish generally spawn at about 2½ inches on a hard substrate and are excellent parents. A family of jewels is a striking sight. The jewel fish is recommended for every cichlid enthusiast. But the five-spot is not recommended for anyone but the most dedicated cichlidophile with a large tank to spare. Both species

This is Hemichromis elongatus, *a species very closely related to the better known* H. fasciatus. *Both species can be treated alike. Photo by Dr. Herbert R. Axelrod.*

pattern, tending to concentric circles. There is no justification for removing the eggs for artificial hatching; the parents will do better than the aquarist.

Breeding the jewel fish

are prolific, usually more so than the aquarist would wish, and many young will probably have to be given away or destroyed.

The *"Pelvicachromis"* Group

The old genus *Pelmatochromis* contained a large number of rather disimilar African fishes, most of them from

Pelvicachromis

Species of the genus Pelvicachromis *normally have a more colorful female — the reverse of the majority of cichlids. This is a female* P. pulcher *(the former* P. kribensis).

One species occurs in East Africa, and there have been arguments to place it in *Haplochromis*. Disregarding this one disputed species, it appears that the typical mode of breeding is of the nesting or substrate-spawning type. Those few species that are now known to be mouthbrooders are placed either in the genus *Chromidotilapia* or in the coastal areas of the western part of the continent. The distribution is generally in fresh or brackish water (lakes, lagoons, rivers) from Sierra Leone to the Congo basin.

A female Pelvicachromis subocellatus. *For spawning these species caves built from rocks or inverted flower pots are recommended. A flower pot with a broken edge for entry can be seen behind the fish. Photo by Juergen Kaspatrick.*

genus *Tilapia.* Most of the well-known aquarium species are now in the genus *Pelvicachromis.*

All *Pelvicachromis* do well in large aquaria (but tolerate smaller quarters at small sizes), with salt, heat, and dense vegetation. A salinity of one teaspoon of salt per two gallons is recommended, and the temperature should be about 80°F for best results.

These criteria, it must be emphasized, are based on knowledge of only a sampling of species, specifically those fishes which have been known to the hobby. Most of the species in this genus have never been aquarium fishes, or even written about in popular literature. The species best known to aquarists is the fish popularly called *kribensis.*

This is a female Pelvicachromis roloffi. *Unfortunately, this species has not become readily available to aquarists. Hopefully this will become rectified in the near future. Photo by E. Roloff.*

Pelvicachromis

The correct name of this fish is now *Pelvicachromis pulcher*. Although *Pelvicachromis* species vary somewhat, they share the presence of a large, fleshy, rough-surfaced pad of tissue medial to the top of the gill arches. A partial breakdown of the group is presented below.

OLD HOBBY NAME	NEW VALID NAME
Pelmatochromis annectens	*Thysia ansorgii*
Pelmatochromis arnoldi	*Thysia ansorgii*
Pelmatochromis kribensis	*Pelvicachromis pulcher*
Pelmatochromis aureocephalus	*Pelvicachromis pulcher*
Pelmatochromis taeniatus	*Pelvicachromis taeniatus*
Pelmatochromis klugei	*Pelvicachromis taeniatus*
Pelmatochromis pulcher	*Pelvicachromis sp. aff. pulcher*
Pelmatochromis camerunensis	*Pelvicachromis sp. aff. pulcher*
Pelmatochromis subocellatus	*Pelvicachromis subocellatus*
Pelmatochromis ocellifer	*Tilapia ocellifer*
Pelmatochromis congicus	*Tilapia congicus*
Pelmatochromis buettikoferi	*Tilapia buettikoferi*
Pelmatochromis guentheri	*Chromidotilapia guentheri*
Pelmatochromis kingsleyae	*Chromidotilapia kingsleyae*

A suggested typical set-up for many non-destructive cichlids. These are cichlids that do not tear up plants or dig extensively. Illustration by John R. Quinn.

The male
Pelvicachromis
pulcher, *however, is
certainly no slouch
either when it
comes to colors.
There are many
different varieties of
these species.*

A male
Pelvicachromis
taeniatus. *Males
lack the bright red
or violet bellies
exhibited by the
females. Photo by
H.-J. Richter.*

A pair of Pelvicachromis taeniatus. *The male is behind the female who is
not showing her full spawning colors as yet. Photo by S. Kochetov.*

Pelvicachromis

The spawning of *Chromidotilapia guentheri* was described by Myrberg in 1965. In this species, the female holds the territory and spawns with a male who enters her domain. There is courtship and mutual acceptance, and then spawning occurs. The male is the egg brooder. Unlike the mbuna of Lake Malawi and other

flees the spawning area as soon as she takes up all

Pelvicachromis pulcher, *male.*

Although Chromidotilapia guentheri *is the species usually seen in the hobby, there are other species that do turn up from time to time — like this* C. kingsleyi. *Photo by Kochetov.*

Certain cichlids will prefer spawning on the roof of caves as seen here in Thysia ansorgii. *Photo by R. Zukal.*

mouthbrooders, in this case the pair remains together. (In the mbuna the female is the brooder and

her fertilized eggs.) In 9 to 12 days the young are released and guarded by both parents. Both parents will take the young in their mouths during periods of danger, and protection of the young extends for 30-40 days post-release (even though the young are too large to hold orally long before this). This method of breeding, pair-bond formation, incubation, protection of the young, etc., shares many similarities with substrate breeders. It most closely resembles the method of *Geophagus jurupari* of South America. The major difference between these two species is that *Geophagus jurupari* waits about 24 hours before orally incubating the eggs, whereas *C. guentheri* picks them up only about 40 minutes after spawning.

As far as is known, most of the *Pelvicachromis* species that have been kept in aquaria in the United States are cave spawners. The caves are generally deep (not downward, but inward), so as to be *shaded* from light. Both parents protect the eggs and young although the female seems to be a better parent. In general, these are shy fishes and should have a choice of caves, many rooted plants, and live foods to initiate spawning. Heat is very desirable. The fish lay large, ovoid, tan eggs, with a light pole distally (away from) and a dark thread proximally (adjacent to the spawning site), by which

Flat rocks are usually advisable for most bottom-spawning cichlids. These are cleaned by the prospective spawners very carefully.

they are attached to a rock wall or roof in the cave.

Species of *Pelvicachromis* are not exactly good parents in aquaria. By this it is meant that there is a high risk that they will eat or ignore their eggs. For this reason, persons interested in saving as many young as possible should use artificial incubation.

Finally, there is good evidence that in some species of the genus *Pelvicachromis,* the percentage of either sex may be correlated with the pH of the spawning tank. The reasons for this are totally unknown. But we do know that at pH 4 to 5 one will get 90% males, and at neutral (7.0) pH 90% females. Thus, one should breed them at about pH 6.5, or else vary the spawning pH over a wide range to assure a balance of sexes. If this is not done, one will have difficulty trying to unload practically all males or all females, and any market demands pairs.

A male Chromidotilapia guentheri *displaying his charms to attract a female to the spawning site. Photo by R. Zukal.*

Congo River Cichlids

A number of bottom-dwelling cichlids have been imported from the Congo, the huge river that drains the entire two countries

Spawning is proceeding with the female depositing her eggs in a pit in the sand. Several pits are dug. Some of them are used after the fry hatch to move them to cleaner quarters. Photo by R. Zukal.

formerly called the Belgian and French Congo. Many smaller rivers empty into this massive stream, which finally reaches the Atlantic Ocean well below the equator at the border of the Congo (now Zaire) and Angola. It is certain that only an extremely small percentage of Congo cichlids has yet been seen by the hobby. As these are mostly bottom-dwellers, it is likely that they come from areas of rapids, much like our own native darters *(Etheostoma* and others).

The first and most popular to come in was *Nanochromis splendens* (formerly called *nudiceps),* found near Leopoldville (Kinshasa) where the Congo spreads out into a lake (the Stanley Pool) before resuming its now short course to the sea. The coloration of *N. splendens* is magnificent, primarily light blue with iridescent green highlights at the front. The female is the more colorful of the two. Many people wonder why their *splendens* don't spawn when they look as though they are about to. The answer is simply that they are not in proper condition. In this species, it is normal for the ovipositor to be partly extruded. *It does not indicate imminent*

A *male* Nanochromis splendens *displaying behind a female of the species. Both are quite attractive, but it is the caudal fin pattern that has attracted the most attention. Photo by H.-J. Richter.*

spawning. Before spawning, the female becomes enormously distended. "As fat as a house" is rather descriptive. Under these conditions, there is no way to turn them off. If you find them this fat in your local shop, buy them, for they will no doubt spawn within a couple of weeks.

Breeding takes place preferably in a cave, and

A closely related species is this Nanochromis parilus. *The checkerboard pattern is missing from the lower portion of the caudal fin in this species. Photo by Gene Wolfsheimer.*

the deeper the better. But they will spawn in a small flowerpot if they are ready and nothing else is available. The eggs are large, ovoid, opaque, white, and hang loosely from very short threads. Both parents defend the eggs and young. I have not succeeded in artificially hatching the eggs (as done with most substrate breeders), but it probably can be done. For a successful breeding, see the report by Richter (1969) in *T.F.H.*

N. dimidiatus is a strikingly lovely fish that came in under various names, such as new "*kribensis*," "*dimidis*," etc. It has been spawned a few times. This is a very peaceful fish, extraordinarily variable in markings and coloration, and should have caves and living foods *in abundance.* Coloration varies from whitish with a horizontal medial black stripe, to darkly orange-brown or orange-red. Females will frequently have white scales above the vent (also seen in *Thysia ansorgii)* or a dark ocellus on the rear of the dorsal fin. The male's lips are usually blood red. Unlike its cousin, *N. splendens,* the females do *not* constantly

show the ovipositor. Without a doubt, this species is destined to equal or surpass *splendens* in popularity.

Perhaps the nastiest of the bottom-dwelling Congo cichlids is *Lamprologus mocquardi.* The generic disposition of this species may be incorrect. In any case, the species looks like

Thysia ansorgii is not a colorful species as can be seen by this male. Aquarists do not seem to have it on their list of favorites and it is not available on a regular basis. Photo by R. Zukal.

a washed-out *splendens*. It can be recognized by a red-black iridescent spot on the gill cover and dark fuzzy markings over the body. Only a single breeding report is known.

Congo River Cichlids

This fish is a cave spawner.

Also from the Stanley Pool area comes the shovelmouth cichlid, *Steatocranus tinanti* (formerly *Leptotilapia* and *Gobiochromis).* This drab-colored bottom hopper has a mouth suited to do two

They are very aggressive about this, but do not pursue the enemy once chased from the particular spot. Their tanks should be covered, as battles between two individuals of this species often end with one fish out of the tank and on the floor.

obvious things: (1) move lots of gravel, and (2) behave as a mouthbrooder. Well, as a matter of fact, *it does neither of these things,* so don't let looks deceive you! It also has an almost human expression to its "face." Its territory includes the entire bottom of the tank, wherever it happens to be at the time.

A third species of Nanochromis *commonly available is this* N. dimidiatus. *It is peaceful and for best results should be provided with the usual caves and live foods. Photo by H.-J. Richter.*

Another species of elongate *Steatocranus* has been coming in recently, but its identification is not presently clear. This is a darker fish, not as belligerent as *S. tinanti,* and with a greenish iridescence on the gill cover (which, on these almost human fishes, looks like a cheek).

Perhaps the most peaceful and easy to breed of the Congo river cichlids is the rather shy lumphead larger it gets the larger becomes the big fleshy pad on its forehead (called a *frontal gibbosity).* These fish spawn in caves and are good parents. Large aquaria are necessary, with heavy planting and many rocks for caves.

The last Congo River

or lionhead, *Steatocranus casuarius.* It is also the largest of the fishes discussed in this section. In markings it resembles the unidentified *Steatocranus* cited above, and shipments of the two species mixed often come in from New York importers. The lionhead achieves a size of about 5 inches in aquaria, and the genus we can discuss is *Teleogramma.* These are all rare fishes in the hobby, worm-shaped *(vermiform),* bottom-hopping cichlids which also spawn in caves. The large, opaque eggs are laid on the roof of a cave, and the parents guard the young. The three species known to aquarists include *T. brichardi,* a rather blackish fish, named in

honor of the Congo collector Pierre Brichard; *T. monogramma;* and *T. gracile.*

The Congo River basin is poor in numbers of species, but it is certain that aquarists haven't seen all of them.

Lamprologus mocquardi *is one of the nastiest of the bottom dwelling Congo cichlids. Photo by Klaus Paysan. Below: The Lumphead or Lionhead Cichlid* (Steatocranus casuarius) *develops a gibbosity on the nape. Photo by H. J. Richter.*

Cichlids of Madagascar

The Malagasy Republic occupies the entire island of Madagascar, which lies in the Indian Ocean, separated from the coast of southeast Africa by the deep Mozambique Channel. A number of species and genera of cichlids are endemic to this tropical island. In 1966, Kiener and Maugé reviewed the cichlids of the island.

The largest genus is *Paretroplus,* with five species: *P. dami, P. polyactis, P. maculatus, P. petiti,* and *P. kieneri.* These are generally deep-bodied fishes with small heads, superficially resembling *Etroplus* or *Cichlasoma centrarchus. P. maculatus* has a large dark blotch above and behind the pectoral fin. *P. polyactis* has short bars along the upper part of the sides. *P. dami* has a rather enlarged snout area.

Paratilapia polleni is a dark, beautiful fish with well developed finnage and a Jack Dempsey front profile. It is covered with light spots set in startling contrast.

Ptychochromis oligacanthus is another pretty species and exists as different geographic races along the island.

They may be dark, light, or intermediate with a series of three or four large blotches. The general coloration is blue-gray to red-brown. The dorsal may be violet, edged in dark red. The caudal and anal may be dark red. The most colorful form is from the area of Mandritsara, in the interior north of the island. The less colorful forms are coastal.

Oxylapia polli is an elongate fish with a dark blotch at the upper angle of the gill cover. It is not an attractively colored species, but its shape makes it easy to identify if you know that it's from Madagascar.

Ptychochromoides betsileanus is generally sunfish-shaped, but it has a very enlarged lump on its forehead, almost (but not quite) as impressive as in the lumphead, *Steatocranus,* from the Congo. It even more closely resembles the African species *Cyphotilapia frontosa.* Again, knowing the origin of the fish, this *frontal gibbosity* makes identification easy.

Tilapia rendalli and *Tilapia mossambica* have been introduced and established on the island.

Apparently none of these native fishes has yet been introduced to the hobby, at least in the United States, so we have something else for cichlidophiles to look forward to. Notice the absence of *Haplochromis* and native *Tilapia* from the island.

Madagascar is a mountainous island rising out of a 12,000 foot sea bed. Due west from the north of the island are the famous Comoro Islands, home of that Devonian relic, the coelacanth.

Hemichromis elongatus.

Tilapia (Sarotherodon) mossambicus.

The Rift Valley System of Africa

The Great Rift Valley is discontinuous. It arises in Northeast Africa, at the southern tip of the Arabian peninsula where the Red Sea meets the Gulf of Aden. It consists of a giant valley running down the midline of a great mountain system to its left and right. At its beginning the west it is also seen, swinging southeast and then southwest and finally straight south. Where there are two great rift valleys, at the disjuncture, a large area between is taken up by the rather shallow, yet enormous, Lake Victoria. This lake does not lie within either of the two rift

Tropheus moorii was one of the early imports from Lake Tanganyika. Since those early days more than two dozen additional morphs have been collected and there are probably still more as yet undiscovered. Photo by Pierre Brichard.

it heads southwest in an arc with the Gulf of Aden, and then abruptly swings to a southeast arc. Far to valleys, but between them. You will have to examine a map to see this. It is usually the lakes along the southern part of the rift valley system that are called the Rift lakes. To find them, locate the mouth of the Nile at Alexandria, Egypt. Follow it south across the Sahara and the Sudan. At about

Many cichlids are diggers. As such they could quickly undermine a set-up such as this one. Rift Lake cichlids do best with a maximum of rocky caves and a minimum of plants.

2°N latitude we find the first Rift Valley lake, Lake Albert. It has a depth of 2,030 feet. From this lake we move south through the Semliki forest along the Semliki River to Lake Edward, almost 3,000 feet deep. The forest and river leading to this lake are in a land called Ruwenzori, (now Uganda) whose southern tip is on the Equator. From Lake Edward below the Equator, we continue our southward journey passing through the eastern limit of the Congo to Lake Kivu with a depth of almost 4,800 feet! The eastern shore of Lake Kivu is the country Rwanda, to the south of Uganda (eastern shore for the other two lakes). Still heading south and overland, we next come to the longest of the rift lakes, Lake Tanganyika. This lake is 4,700 feet deep and covers an area of 12,700 square miles. It is bordered by the Congo to the west, Zambia to the southwest, Tanzania to the east, and Burundi to the northeast. Trekking overland to the east-southeast, we pass the relatively insignificant Lake Rukwa, and then head south where we come upon Lake Malawi (formerly Lake Nyasa). Its southern and western shores are the Republic of Malawi (formerly Nyasaland), and its eastern shores are Tanzania and then Mozambique. Lake Malawi is 2,600 feet deep and occupies 11,000 square miles. South of it, lodged between Malawi and Mozambique, is Lake Chilwa. To the west of the Rift Valley Lakes are Lakes Mweru and Bangweulu. There are many other major lakes along and throughout East Africa,

Contrary to almost every mbuna species, the male of Pseudotropheus lombardoi *is the upper gold colored fish and the blue one is the female. Photo by G. Meola, African Fish Imports.*

and they are virtually unknown to the aquarium hobby.

At present, aquarists are most familiar with some of the fishes of Lakes Malawi and Tanganyika, but no doubt they will become more familiar with fishes of the other lakes in time, if and when those lakes become more open to fish exporters. Whereas the principal Rift Valley lakes (Albert, Edward, Kivu, Tanganyika, and Malawi) are deep clefts in the earth's crust, Lake Victoria is generally considered a saucer, only 270 feet deep, but 26,000 square miles in area!

Lake Malawi

Although a very deep lake, fish life is restricted to the upper waters, where the water is oxygenated and varies from 74 to 82°F (about room temperature) all year round. The deeper waters of the lake are slightly cooler, and very low or lacking in oxygen. The inflow from rivers is insignificant, and the lake receives most of its water from rain, usually in the form of night-time prolonged and violent thunderstorms. Beauchamp in 1964 described this and other rift lakes in considerable detail. Much of the surrounding region is arid.

Because land temperatures fluctuate considerably, but not the water temperature, the mornings consist of warm air rising over the lake. This is actually visible, as these warm columns may be occupied by swarms of

Undergravel filters are all right for some cichlids. The diggers, however, will soon expose the filter making it useless.

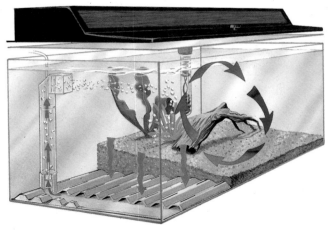

Julies from Lake Tanganyika also burst upon the scene and became instant favorites. This Julidochromis ornatus was one of the first species imported. Photo by H.-J. Richter.

flies looking, as Beauchamp has said, like "bush fires several miles out on the lake." This situation applies to the other rift lakes as well. As the warm air over the lakes rises, it creates strong winds heading out to the lakes. The African name for Lake Albert means "destroyer of locusts," as a consequence of the winds blowing the locusts out to the lake where they eventually fall and drown. The low temperature variation of the lakes prevents the lakes from "turning over" annually as do the American Great Lakes. Turnover would bring up nutrients from the bottom and oxygenate the depths, but this doesn't happen in the deep rift lakes. Instead, organic material continues to fall to the bottom where it accumulates as a gray or black mud.

A group of beautiful fishes is found along the rocky shores of Lake Malawi. These are all recently evolved from a *Haplochromis* type of ancestor, and the natives refer to the group as mbuna (pronounced EM-BOO'-NUH).

71

A variety of cichlids. Due to their aggressiveness cichlids sharing the same tank must be chosen with utmost care. Check with your pet store manager to see which species are compatible.

Breeding the mbuna or *Haplochromis*-derivatives presents no difficulty. So far a number of them have been imported and bred by aquarists all around the country. Because the price

The Red-top Labeotropheus (Labeotropheus trewavasae). The snout protrudes like this in both species known to the genus. In successive generations of aquarium bred fish this snout seems to become smaller. Photo by Andre Roth.

of some of the more spectacular of these fishes still remains in the range of twenty to seventy dollars a pair for adults, many aquarists find it wise to purchase a number of young of one species, raise them, breed them, and trade fry through the mails.

The dominant male *Pseudotropheus* or *Melanochromis* (usually the biggest) develops the best colors, although in *Melanochromis auratus* almost all males show good color, and digs out a clearing in the gravel, usually down to the floor of the aquarium. When a female approaches, his colors intensify and he darts out to display to her. The display consists of swimming around her, back and forth, cutting off her avenues of retreat, all the while with fins spread wide. If the female is in condition (quite fat, sometimes with a nipple where other cichlids develop a distinct spawning tube), her color will be muddy in stark

73

A strain of albino Pseudotropheus zebra *has been developed. As you can see they are quite attractive. They should have rocks formed into caves rather than plants, however. Photo by Ken Miner.*

The Golden Cichlid (Melanochromis auratus) *is a very aggressive species. Breeding males and females are easy to distinguish with males (lower fish in photo) taking on a reverse color from the females (upper fish in photo).*

contrast to the male. Spawning commences with the female brushing her vent across the "nest" and leaping out of the way as the male does the same. Then he jumps away as she again passes her vent over the nest. While the male is fertilizing the eggs, the female is picking them up in her mouth as fast as she can. The eggs are very large, yellowish white, millet-shaped and non-sticky, and you can see them being blown around by the currents set up by the fish's activity. Often the female picks them up while they are blowing about in the water above the nest. The entire spawning vibrates his body rapidly, fins erect, both prior to and during his expulsion of sperm. The female doesn't vibrate to anywhere near the same degree, if at all. At the conclusion of the spawning, the female flees the area, as the male still wants to continue. She,

Cichlids are normally not good pond inhabitants. Some of them (especially Tilapia *species), however, can be bred there if the water is kept warm enough.*

however, has had it and takes off for distant parts, usually among floating plants at the other end of the tank. She should now be removed to a separate tank, or the male should be removed. This depends, of course, on whether they have spawned in a private aquarium or in a

usually lasts in the neighborhood of 30 to 45 minutes or more. In general, the excited male

This female Pseudotropheus "dingani" is brooding a batch of eggs. Such females are best placed in a quiet tank of their own for the approximately three weeks it takes. Photo by Dr. Herbert R. Axelrod.

necessary to feed the female. Some will not eat at all (*P. zebra*), some will spit out the eggs to eat and then pick them up again (*M. auratus*), and some will carefully eat with the eggs still in the mouth (lavender cichlid). This variation among species probably applies within species as well, so don't be surprised if your fish do it differently.

Toward the end of the incubation period, the community tank, and both methods are used about equally by different aquarists. There is no danger in netting out the female. She will rarely, if ever, spit out or swallow the eggs.

The incubation period varies from three to five weeks, according to temperature and species. During this time it is not

This female Pseudotropheus microstoma *has released her fry so that they can forage for themselves. Some of the fry are trying to return to the safety of her mouth. Photo by Uwe Werner.*

female's throat becomes blackened as a consequence of the babies inside. In the last days, she will search out the thickest plants or mops in anticipation of the blessed event. She may release all her fry at once, or she may spit out a few a day, picking them up again at night or if they still have yolk sacs. There is much variation at this stage. She warning signal to other fishes to keep away. Eventually all the fry are out to stay and the female should be removed. But you aren't done with her yet. She should not be placed back in the community tank in this weakened condition, but should be placed in a recuperation tank and fed the very best foods. When she has regained her

This mbuna has had problems with its name. Most aquarists know it under the name Pseudotropheus "eduardi". The black trim is absent in another variety. Photo by Bernard Brymiller.

develops a pattern resembling that of a male, but not nearly so intense, and this is probably a strength (about two weeks), she may be placed in the community tank once again. In the meantime, the fry should be fed brine shrimp in quantity and moved to larger quarters as necessary. Spawns vary from a few to fifty or more, depending on the size of the female more than

anything else.

There are a few principles to keep in mind with the Malawi community tank. First, the water should never be allowed to become acid. It is best to keep it alkaline, hard, and somewhat salty. It should be clean and well aerated. Aeration is also of major importance to the incubating female. In dirty or stagnant water it is quite common for them to spit out the eggs. It also seems likely that the eggs that were spat out were dead, probably due to the high bacterial count in dirty water.

These fishes seem to breed more frequently if there are a number of the

One of the more colorful Lake Tanganyika cichlids is this Lamprologus leleupi. Unfortunately, the brilliant yellow color seems to fade in domestic strains. Photo by H.-J. Richter.

same species in the same tank, rather than just a pair. A fresh water change often triggers excitement, leading to the female deciding to go ahead after all. Large tanks are preferred, as these are all very active fishes. Exceptions to keeping numerous males of one species include *M. auratus* and *P. elongatus,* both species being quite belligerent to their own kind. Males of most other species get along much better.

Sexing is frequently difficult. In many cases, both sexes are marked identically when not in nuptial colors. Spots on the anal fin ("egg spots") are not a sure indication of sex.

The species most often seen today include *Pseudotropheus zebra, P. tropheops, P. fuscus, P. novemfasciatus, P. elongatus, Labeotropheus fuelleborni, L. trewavasae, Labidochromis vermivorus,* as well as numerous *Haplochromis*

species and peacocks (*Aulonocara*).

Zoologically, Lake Tanganyika also has some beauties, including *Tropheus moorii, T. duboisi, Julidochromis* species, *Lamprologus compressiceps,* and *L. leloupi.* The Lake Tanganyika fishes seem to be mostly bottom spawners, but *Tropheus* species are mouthbrooders. At the moment numerous color varieties of *Tropheus* are being both imported and bred, literally dozens of large and small *Lamprologus* species are available, and various oddballs such as goby cichlids (*Eretmodus, Spathodus*) are not uncommon. Many new species are still coming in from Lake Tanganyika exporters, and the variety of fishes available may eventually exceed the number coming from Lake Malawi.

An African lake cichlid tank should have very hard and alkaline water. Its decorations should consist mainly of rocks and perhaps a bit of driftwood as shown here. Photo by Dr. Herbert R. Axelrod.

The most popular Lake Tanganyika cichlids currently are the shell dwellers. This one is Lamprologus brevis which lives and breeds in the shell seen in the photo. Photo by H.-J. Richter.

American Cichlids

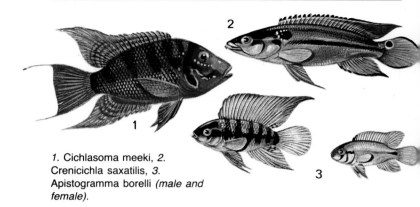

1. Cichlasoma meeki, 2.
Crenicichla saxatilis, 3.
Apistogramma borelli *(male and
female).*

Cichlids in the New World are found from Texas, along the eastern and southern coasts of Mexico, to Central and most of South America and some Caribbean islands. The lone native U.S. species is *Cichlasoma cyanoguttatum,* a large and beautiful animal found in the same water with centrarchids (the basses and sunfishes). Exotics have been introduced into various American waters. Some dig pits in sand or gravel, others clean a hard surface for spawning, and others may subsequently pick up the eggs for mouthbrooding after a long exposure to the open terrain habitat. Except for

*The only cichlid native to any part of the United States is the Texas Cichlid (*Cichlasoma cyanoguttatum*). It breeds well and is a popular species with aquarists.*

these belated mouthbrooders *(Geophagus jurupari* and perhaps some others in the genus) and the fishes we know little or nothing about, all the American cichlids then fall into two categories of interest to the breeder: pit spawners (in which case one usually cannot artificially incubate the eggs) and hard substrate spawners. The hard substrate may be a flowerpot, rock, strip of slate, or the leaf of a large real or plastic plant, e.g., *Echinodorus* spp. For such fishes artificial incubation is usually possible and more successful than natural incubation by the parents. An exception is *Symphysodon discus,* in which the fry are heavily dependent on parental slime secretion.

Geophagus jurupari *is commonly referred to as an "earth-eater". It picks up bottom material and sifts through it for edible items. This can be very destructive to a tank's decor. Photo by H. Hansen.*

The Genus *Nannacara*

N. anomala is easily recognized; this species was reported from western Guyana and Venezuela. Popularly called the golden-eye dwarf cichlid, a better name might be the slate or gray dwarf cichlid to indicate the usual color of the male. Males attain larger size than females, usually over two inches, although larger than three inches is not unusual. Females are considerably smaller, average adult size being under 1½ inches. They may be kept in their own five gallon tank or in a community tank, and will spawn with equal likelihood in both places. Prior to spawning the male develops his best nuptial

Nannacara anomala *(male and female).*

colors and is rather rough on the female. After spawning, which occurs in a rock cave, flowerpot, or on a rock out in the open (if nothing else is available) the tables are turned. The female becomes the dangerous partner and the male is driven off. In a small aquarium he may be badly beaten by his diminutive mate and to its having happened is the pattern of the female. After spawning has taken place, she develops a marked checkerboard pattern and spends most of her time with the eggs, coming out occasionally to see that the vicinity is clear of hostiles. The eggs hatch in 2 to 3 days, and the fry are free-swimming in an additional five days. They

should be removed. The female takes care of the eggs and fry and may be considered an excellent parent. If spawning occurs in a community tank, the male is safe, but the fry will gradually be snapped up one by one by the other tankmates despite the best efforts of the female. Thus, it is best to remove the eggs for artificial hatching. In tanks where it is difficult to tell whether spawning has occurred, the best clue

Certain cichlids can be made a part of a community tank. Here Angelfish and Guppies are housed together. Remember, however, that the Guppy fry will be eaten by the Angelfish. Photo courtesy of Werther Paccagnella.

should be started on newly hatched brine shrimp and microworms. The fry behave as a shoal, and if the female is left with them she will herd them carefully to and from feeding and resting areas. Juvenile females of this species will

This is a male Nannacara anomala *in full breeding colors. The drab female is just below him. As a small, peaceful species it can be recommended for novice aquarists. Photo by Burkhard Kahl.*

also herd and protect groups of fry of unrelated cichlids, including *Apistogramma* and *Pelvicachromis.*

The Genus *Apistogramma*

This large genus of usually difficult-to-identify species ranges through most of South America. Some are usually pit spawners (e.g.,

A. ramirezi), but most are hard surface spawners. Their identification is a headache even for ichthyologists. Only a few species are easily recognized.

Most *Apistogramma* species require warmth, clean water, and privacy. It is best to set up pairs in 7 gallon aquaria. The tanks should be in a heated room or contain heaters; there should be sand or gravel (preferably sand) in the tank, some rooted plants, and a flowerpot or facsimile. Undergravel filtration is desirable.

As these fishes are not

the best of parents, it is advisable to handle the first couple of spawnings artificially to assure that you have some offspring. Spawnings run to around 150 eggs, but one raises about 30 to 50 fish after all the mortality, most of which occurs among the eggs. Some species are ready and prolific spawners (*pertense, ramirezi*), whereas others are not as quick to spawn (*agassizi*), although they will—seemingly grudgingly! All are susceptible to bacterial infections in dirty water. Thus, keep the sand surface clean, the water clear, warm, and well aerated, and the plants lighted and healthy. If the tank looks like part of a forest brook, the fish should look good too. These fishes will not tolerate laxity in aquarium management.

The ram, *ramirezi,* is usually considered intermediate between true *Apistogramma* and *Geophagus* or *Aequidens.* It is often placed in its own genus, *Microgeophagus.*

The Genus *Geophagus*
This has been a confused genus for many years, at one time containing species now in *Apistogramma* or

Male Apistos are almost always much more colorful than the females. This male may be Apistogramma bitaeniata but the dorsal fin seems to be too tall for that species. Photo by Dr. J. Vierke.

elsewhere. The distribution ranges from Panama in Central American to Argentina in southern South America. The genus is characterized by the form of the gill arches, by

One of the many species of Apistogramma is A. viejita. The red borders to the caudal fin help identify it. Photo by H.-J. Richter.

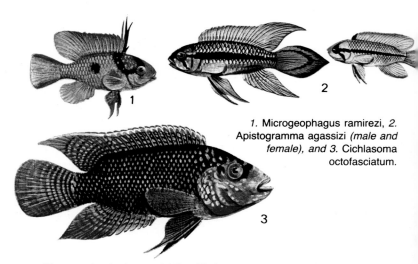

1. Microgeophagus ramirezi, 2. Apistogramma agassizi *(male and female), and* 3. Cichlasoma octofasciatum.

The spotting in the caudal fin of Apistogramma cacatuoides *is very variable. It can be well developed as in this individual or entirely absent. Photo by Dr. Herbert R. Axelrod.*

Apistogramma trifasciatum has been around for quite some time. The third stripe (behind the pectoral fins) is distinctive. Photo by H. Hansen, Aquarium Berlin.

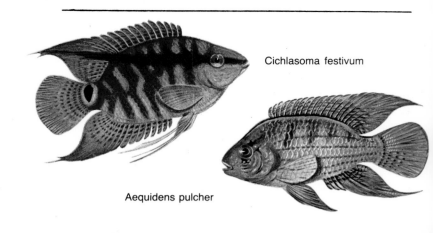

Cichlasoma festivum

Aequidens pulcher

tne distance between the lateral line and the dorsal fin, and by the end of the lateral line, where it enters the caudal, being divided (bifurcate). These fishes are all diggers, and have well-shaped snouts for that particular purpose. They are medium to large cichlids and should have large aquaria with much coarse gravel. Outside filtration is preferred.

The most popular species is *G. jurupari,* but much of what has been written about it may refer

Perhaps the best known dwarf cichlid is the Ram, even though it may be called Apistogramma ramirezi, Papiliochromis ramirezi, *or the currently accepted name* Microgeophagus ramirezi. *Photo by H.-J. Richter.*

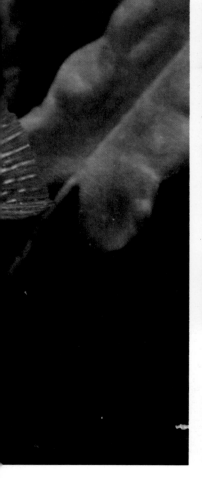

Species of Geophagus *have also recently received a surge of interest. This attractive* Geophagus australis *is one of the species that is responsible. Photo of a male by Aaron Norman.*

to the very similar *G. acuticeps,* which has long, pointy ventrals. The next most popular species is *G. brasiliensis.* Other species include *G. australe, G. gymnogenys,* and *G. surinamensis.* Closely related to *Geophagus* is *Biotodoma,* with one of its two species, *B. cupido,* in the hobby.

Except for *G. jurupari, G.*

This unusual species is currently known as Geophagus steindachneri *although another name,* G. hondae, *is still in use. Photo by H.-J. Richter.*

Geophagus rhabdotus is often listed under the name Gymnogeophagus rhabdotus. *However, it is still the same fish. Photo by Uwe Werner.*

surinamensis, and *B. cupido,* all the others may well be substrate breeders. *B. cupido* is said to be a mouthbrooder, but no breeding reports are known in the hobby literature in this country. *G. jurupari* begins as a substrate breeder in the usual sense, but about 24 hours after spawning both

93

Cichlasoma severum

parents pick up the eggs for oral incubation. This incubation period lasts about eight days. However, the parents may accept the young into their mouths during periods of danger for up to 30 days.

This semi-albino (without pink eyes) is called the Pink Convict Cichlid in the hobby. Photo *by R. Zukal.*

The Genus *Cichlasoma*

This is one of the largest genera of American cichlids and contains a number of old aquarium favorites.

Most of these species are Central American, and the remainder (except for the one from the U.S.A.) are from farther south. They occupy habitats in nature similar to the habitats occupied by our native sunfishes.

Commonly available in the hobby are an albino strain of *C. severum,* as well as a semi-albino (black eyes) of the same species, and a semi-albino strain of *C. nigrofasciatum,* commonly called the "pink convict." This last was developed by Ken Grisham of Fort Worth. This fish is now, probably, the most popular *Cichlasoma* in the hobby, as it thrives in Florida's hard, alkaline water, and the big fish farms are mostly in Florida.

Most of the species are hard substrate spawners, but some spawn in sand or gravel pits. In general, males and females are similarly marked, but males tend to have longer and more pointed dorsal and anal fins. A strong pair bond is established between compatible fish, and bond formation may involve display, snapping, jaw-locking, and other forms of pseudo-belligerent behavior. Generally, both parents clean the spawning site of all debris and algae, and the ends of the genitalia (genital papillae) can be observed on both parents a day or more before spawning. In the female it is blunt and in the male it is very fine and curved.

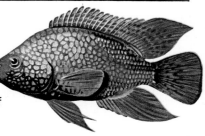

Cichlasoma carpintis

The spawning may vary from 100 to 1,000 eggs, depending on the size of the parents. The territory is vigorously defended. Both parents clean the eggs and wash the newly hatched fry in their mouths. Fry, after hatching, are generally moved to one or more pits, in turn, apparently because it is easier to make up a new pit than to keep an old

one clean. There are all degrees of parental care, and all degrees of ease of spawning in aquaria. For the better species, it is wise to employ artificial incubation of the eggs for the first spawning or two, i.e., until you are assured that you will have all the offspring you want.

The double striping is unique to this Geophagus balzanii. *The male develops a nuchal hump which can grow much larger than the one seen here. Photo by H.-J. Richter.*

One of the newer favorites of the genus is Cichlasoma synspilum. Aquarists spawning this species have a ready market for the young. Photo by Rainer Stawikowski.

This is a colorful male Chocolate Cichlid (Cichlasoma coryphaenoides).
It is aggressive and often hard to keep but still has its devotees. Photo
by Aaron Norman.

The Black Belt Cichlid (Cichlasoma maculicauda) received its common
name not from its aggressive tendencies but from its color pattern.
Photo by Ken Lucas, Steinhart Aquarium.

The Genus *Aequidens*

The Genus *Aequidens*

The genus *Aequidens* Eigenmann and Bray is characterized by gills without a lobe, small gill rakers, small and moderately protractile mouth, low lateral line, and the lateral line scales being the same size as the other scales. There are a large number of species in this genus, occurring from Central to South America.

Aequidens species are not generally plant uprooters, and most species can be kept in

tanks with large rooted Amazon sword plants. They are hard surface

Aequidens curviceps *is a small, peaceful, easily spawned species. It is a good beginner's cichlid that can even be housed in a community aquarium. Photo by H.-J. Richter.*

spawners and should be provided with rocks and large caves. Only the largest of them may be considered diggers, such as ports. Years ago, the port (*A. portalegrensis*) was considered one of the easiest cichlids to breed

Genera *Pterophyllum* and *Symphysodon*

and was extremely popular with aquarists. Today, it is hardly ever seen in shops. Another name for this fish is the "black acara."

Young specimens of *Aequidens* are often difficult to identify, and the aquarist purchasing young fish is really taking pot luck. When grown, however, a number of species are rather distinctive and may be identified from good aquarium photographs.

Spawning follows the usual situation as described for *Cichlasoma.* Eggs are laid on a hard surface and should be removed for artificial incubation, except in *A. portalegrensis,* which is an exceptionally good parent. Most species do not spawn as readily as species of *Cichlasoma,* and most are not as hardy. Clean water is required. They should not be mixed with very rough fishes or are likely to suffer.

Genera *Pterophyllum* and *Symphysodon*

The common angelfish (*Pterophyllum scalare*) and the common discus fishes (*Symphysodon discus* and *S. aequifasciata*) are native to Amazonian waters of South America, especially deep quiet waters, where they may be found with other such quiet-loving cichlids as *Cichlasoma festivum* and *C. severum*. Both angels and discus are laterally compressed, deep-bodied fishes well suited for gently gliding through tall underwater grasses and lilies.

Magnificent green discus developed by Dr. Schmidt-Focke from the Tefe samples of Symphysodon aequifasciata aequifasciata collected by Dr. Herbert R. Axelrod. Photo by A. L. Pieter.

Young Symphysodon aequifasciata *usually exhibit the vertical stripes inherent in the species. Their lower price usually makes them available to most aquarists.*

Symphysodon aequifasciata

Spawning is not typically substrate, because the substrate is off the bottom and out in the open. In aquaria, eggs are laid on the leaves of large plants or on a strip of slate angled against the side of the tank. Both species are usually very peaceful, and only occasionally aggressive. Territories are

*Many varieties of Discus (*Symphysodon aequifasciata*) are currently available ranging from solid blues to almost all brown. This one has most of the coloring around the periphery of the fish.*

103

One of the colorful strains of Symphysodon aequifasciata with emphasis on the red coloration.

only staked out at spawning time.

I know of no sure way to sex these fishes morphologically, and most aquarists who are successful breeders do

Aquarists call this fish Heckel's Discus. It is actually Symphysodon discus Heckel and is characterized by the heavy dark central band.

The gold background color and the bright red eye complement nicely the aqua coloring of this Discus. Note the barring of Symphysodon aequifasciata *is still obvious.*

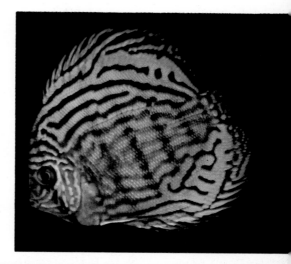

their sexing by behavior. Angels are easy to breed and raise, especially

This is Dr. Schmidt-Focke's "full blue" Discus. The blue coloration is no longer restricted to lines across the body but covers it entirely. Photo by Hans Mayland.

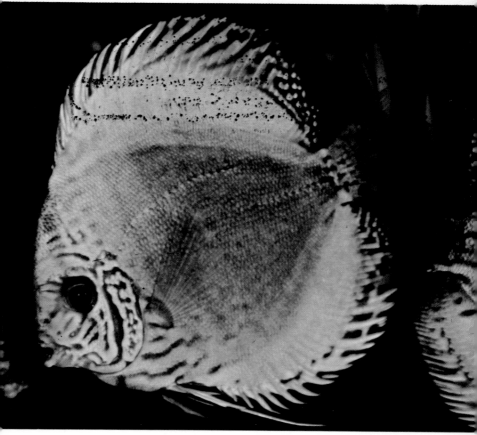

One fish that is not seen very often is a wild Angelfish (Pterophyllum scalare). This example was collected and photographed by Dr. Herbert Axelrod.

Model of an Angelfish. Photo by Dr. Herbert R. Axelrod.

Selective breeding has developed numerous Angelfish strains. This is a Veiltail Marbled Angel. Photo by Andre Roth.

artificially incubating the eggs. If one goes the artificial route, the parents will spawn more frequently. But the aquarist who wants to see a magnificent sight is advised to let the parents raise the fry. The discus fish is difficult to maintain in good health, more difficult to spawn, and even more difficult to raise.

One Florida breeder rears the fry away from the parents, but in discus fish the average aquarist must let the parents raise the fry. The reason is that the fry are heavily dependent on the slime secreted by the parental fish as a first food. This was first brought to the attention of the hobby by Wolfsheimer and has since been intensively studied by scientists. See also the section on Asian cichlids.

The Black Lace Veiltail Angelfish is hardier than the true black and to some people much more beautiful. Photo by Andre Roth.

Pterophyllum scalare

A magnificant fully adult Golden Angelfish. Some Gold Angels do not develop their golden color immediately but change as they grow. Photo by H.-J. Richter.

An interesting view of young Discus gathering about one of the parents. This is not only for protection but they also feed off the secretions produced by their parents. Photo by Petersmann.

Astronotus and Miscellaneous Genera

Astronotus and Miscellaneous Genera

Ranging in nature from eastern Venezuela, through the Amazon Basin, to Paraguay, *Astronotus ocellatus* is one of the largest fishes generally kept in home aquariums. Reaching almost a foot in length, with an appetite to match, the oscar requires a tank of about 50 gallon size when mature. The more colorful juveniles are preferred by most hobbyists.

A number of cichlid genera are relatively unknown to aquarists, but their popularity should be anticipated as the demand for less commonly kept fishes continues to grow.

Chaetobranchus Heckel contains two crappie-shaped yellowish species, *C. flavescens* and *C. bitaeniatus*. Neither has yet been reported as bred. *C. bitaeniatus* is said to be peaceful and not a digger.

The genus *Acaronia* (synonym: *Acaropsis*)

*A fully adult Oscar (*Astronotus ocellatus*) with a normal wild color pattern. Several different strains, including a relatively new long-finned version, are available. Photo by Klaus Paysan.*

contains two species, *A. nassa* and *A. trimaculata*. This genus is very close to *Aequidens*. They have not yet been reported as bred.

A pair of Herotilapia multispinosa surrounded by literally hundreds of fry. The young are guarded by both parents for several weeks. Photo by H.-J. Richter.

Acaronia nassa *is a nasty, pugnacious cichlid that has a large appetite. It usually is not imported for itself but turns up as strays in shipments of other species. Photo by Harald Schultz.*

Uaru Heckel contains two species, *U. amphiacanthoides* and *U. imperialis.* They are said to spawn like angels, but the young are difficult to raise. Perhaps parental slime is required. Young *Uaru amphiacanthoids* are occasionally seen for sale.

Herotilapia contains one species, *H. multispinosa.* This is the "rainbow cichlid." It differs from *Cichlasoma* in having tricuspid teeth (whereas *Cichlasoma* has conical teeth). A very easy fish to breed and raise, it may rival the congo or convict cichlid in popularity among beginning cichlidophiles. Although attaining about four inches, they will breed at 1½ inches.

Acarichthys Eigenmann may contain one species, *A. heckeli.* This genus may prove to be a junior synonym for *Geophagus.* It is greatly admired for the highly filamentous fins of the adult.

Neetroplus nematopus is from the Atlantic drainage of Costa Rica and also occurs in Lake Nicaragua. It is not uncommonly available.

There are two pike-like genera, *Batrachops* and *Crenicichla.* In the former genus, the inner teeth are not depressible. *Batrachops* contains *B.*

Astronotus ocellatus

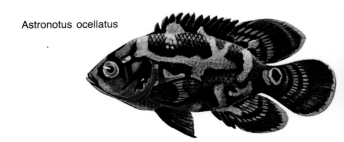

Acarichthys heckeli *is a large, beautiful fish. It is very much like a species of* Geophagus *and requires similar care. Photo by Harald Schultz.*

*The Little Lake Cichlid (*Neetroplus nematopus*) is generally peaceful, except when spawning. It does best in water that is on the hard, alkaline side. Photo by H. Ross Brock.*

cyanonotus, *B.*
nemopterus, and *B.*
reticulatus.

Crenicichla contains a
number of species. These
appear to be pit spawners,
with the male taking care
of the spawn. Generally
rough fishes, preferring
lots of room and lots of
plants for hiding, in nature,
many travel in schools.

Crenicara maculata
bridges the gap between
the dwarfs and the regular-
sized cichlids. Peaceful,
very distinctive, and not
rare, these cave-spawners
sometimes are found in
shipments of mixed dwarf

Crenicichla saxatilis *is a substrate
spawner. In this pair the female is
in front. In this pose the female
reminds one of some of the
females of the genus*
Pelvicachromis. *Photo by
Kochetov.*

The Checkerboard Lyretail Cichlid (Crenicara filamentosa) *is a dwarf cichlid that is peaceful with other species but quarrel among themselves. Photo by Aaron Norman.*

cichlids; rarely are they correctly called "checkerboard cichlids."

The genus *Petenia* Günther was named for the state of Peten in northern Guatemala, at the base of the Yucatan Peninsula. It contains one species, *P. spectabile.*

Some of the larger cichlids such as Astronotus *can be very messy in their eating habits, creating much organic debris to swirl around in their tank. Power filters can be especially useful with such species.*

Artificial Incubation of Eggs

Item one is: get your notebook. When a spawning is observed, check the pH, hardness, temperature, quality and quantity of light, and record these data. Other notes should also be made of the parents, colors, patterns, where and when they spawned, number and color of eggs, approximate size and shape, whether they have visible threads, etc.

Fill a clean gallon jar or two-gallon drum bowl with filtered water from the breeding aquarium. Either take advantage of an outside filter, if it is hooked up to the tank, or filter the water through a cloth net, such as one used to strain brine shrimp. Lacking these, use a well-washed handkerchief.

Grasp the rock or leaf containing the eggs and swirl it gently in the breeding aquarium to remove the bits of debris. Lift it out of the water and place it in the jar of filtered tank water. The eggs should be hanging downward, so that the fry, upon hatching, can drop away from the patch (which may contain fungused areas). Place an airstone at the bottom of the jar and play a gentle stream of air nearby, but not directly on, the eggs. If the eggs are hanging by filaments, adjust the air so

Most cichlids (like this Cichlasoma nigrofasciatum) *will attack any potential threat to their eggs, even something as large as a person's finger. But some individuals for a multitude of reasons may eat their own eggs. For this reason artificial incubation of the eggs is necessary. Photo by R. Zukal.*

Removing eggs from the aquarium may at times be difficult. When they are deposited on a slate, small stone, or even a leaf (as seen here), they can be transferred to a hatching tank with a minimum of trouble.

that those closest to the airstream gently bob in the current.

Add liquid methylene blue until the water is so dark that you can hardly

see your hand behind the jar. *Powdered dye should never be added to the brooding jar.* As an alternative to methylene blue, you may use a weak solution of acriflavine (a favorite of killifish people) or a dilute solution of malachite green (used in trout hatcheries). A combination of dyes may be used, but if so, then the methylene blue should be added last. In general, methylene blue is antibacterial and the other two dyes exert their action on fungi. But there is overlap of activities of all of them.

If a flowerpot, slate bar, piece of shale, or piece of petrified wood contains the eggs, it will absorb much of the dye in a day or two. More dye should be added as needed, until hatching.

The jar containing the eggs, dye, etc., is placed inside a large paper bag (supermarket type), and allowed to sit on a table or high shelf. If left on the floor, the temperature may be too low and the eggs may be killed. The bag protects the eggs from bright light, which *may* (we don't know for sure) be harmful to them.

The eggs are checked daily for progress. If fungus patches appear, these should be removed with an eyedropper. If the patches become too great, squirt the eggs off the rock onto the floor of the jar, and remove the rock with the fungused eggs. Keep the jar as clean as possible. After about 3-7 days, eggs will have hatched and the babies will be wiggling their tails furiously on the floor of the jar. In another few days, their yolk sacs will be considerably reduced, and the fry will remain upright on the floor of the jar, scooting about in circles. In another day or two, they will be able to swim, and may now be transferred into a five or ten gallon tank with clean water from the parental tank. A bare tank is best.

It is surprising how fast fungus can attack a batch of eggs. Fungused eggs should be removed promptly (the spawning pair tend to this if left with the eggs) and/or a suitable fungicide added. Otherwise, your eggs might wind up looking like this. Photo by H.-J. Richter.

Care of Cichlid Fry

During these early days, the fry will appear to have little horns on their heads. These are mucous glands, and the fry will stick to various surfaces — and each other — by means of these head glands. In a few more days, they disappear. Do not feed until the yolk sac is practically gone and the fry appear to be picking on the floor of the tank. Start the fry on microworms or, if large enough, on brine shrimp nauplii. Feed live foods only. Maintain the fry on live baby brine shrimp until they are easily large enough to take prepared foods, such as blended meats. Avoid the use of dried foods. It is important to feed the fry often on live foods during the early days to get them off to a strong start. After they have attained at least a half

Newly hatched fry (in this case Lamprologus leleupi) *usually have a sufficient amount of nutrition in their yolk sacs to last them a few days. Be sure to have proper sized food ready for them when they have to depend upon outside sources. Photo by H.-J. Richter.*

inch, you may taper off, feed less often, and feed cheaper foods (blended meats).

The fry should be transferred to larger quarters long before they seem to require those quarters. This is to speed maximal growth. A large surface to volume ratio is

Care of Cichlid Fry

of considerable importance, and many breeders use special growth tanks; these are generally large and very low, giving a maximum surface area. One can construct growth tanks with wooden frames, chicken wire, and sheet plastic (which comes in large rolls, found in building supply stores). Be sure no nails or wire ends come in sharp contact with the plastic. Heavy aeration is desirable when the fry are large.

For those species which include much vegetable matter in their diets, add

It is always a question as to whether to leave the young with the parents for a period of time. It is very rewarding to be able to observe a "family" like this Pelvicachromis taeniatus *doing what comes naturally. Photo by H.-J. Richter.*

some hair algae, canned spinach, or cooked water sprite or duckweed. Rinse well before adding cooked vegetable material to the fry tank, and use in only slight amounts, replacing it as it is used. Keeping a light on for 24 hours a day will result in maximal growth.

Index

One of the most popular of the cichlids is the Angelfish, Pterophyllum scalare. This is a domestic variety called the Marbled Angelfish. Photo by Andre Roth.

A COMPLETE INTRODUCTION TO

CICHLIDS

CO-011

Cichlasoma octofasciatum grows relatively big and a suitably large tank must be provided. Photo by Dr. Herbert R. Axelrod.